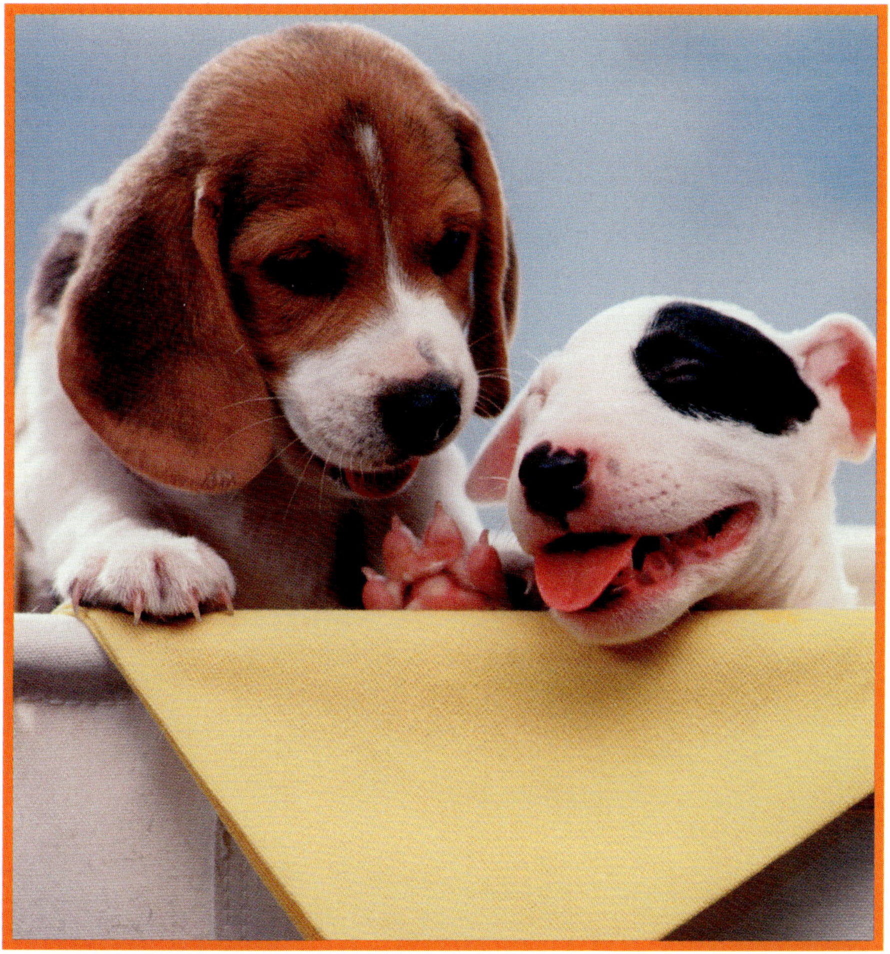

Yip! Yap!

Pups!

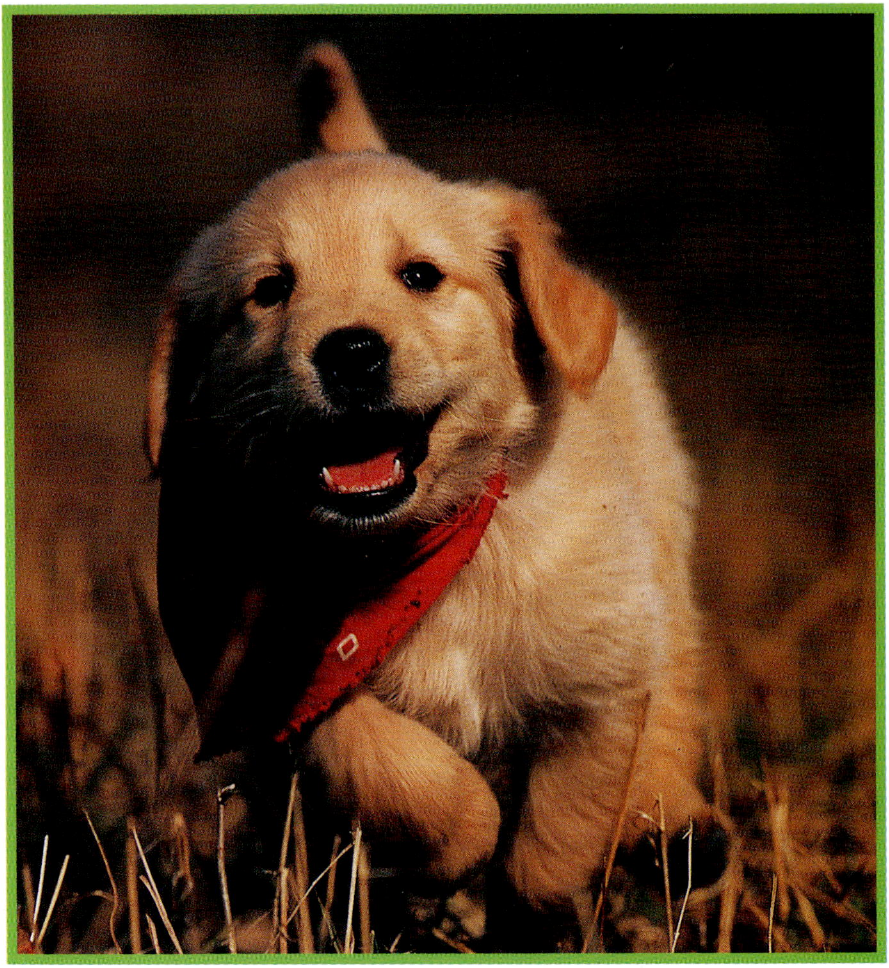

Pups run to you.

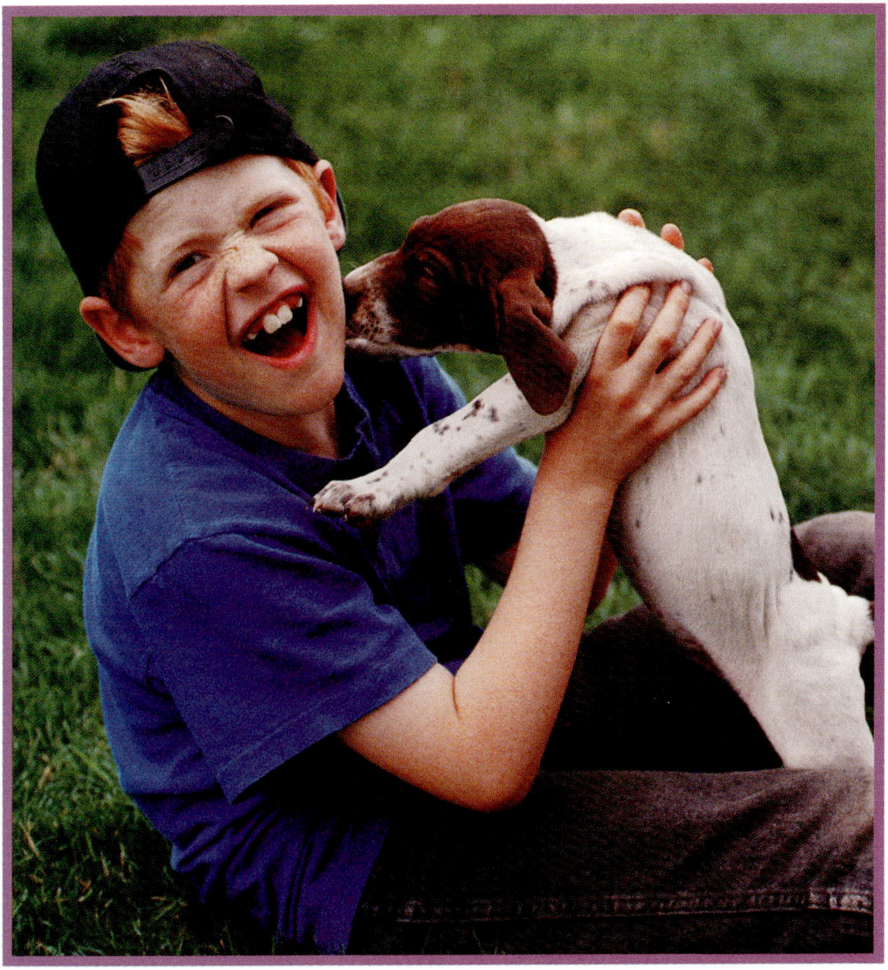

A pup can kiss you.

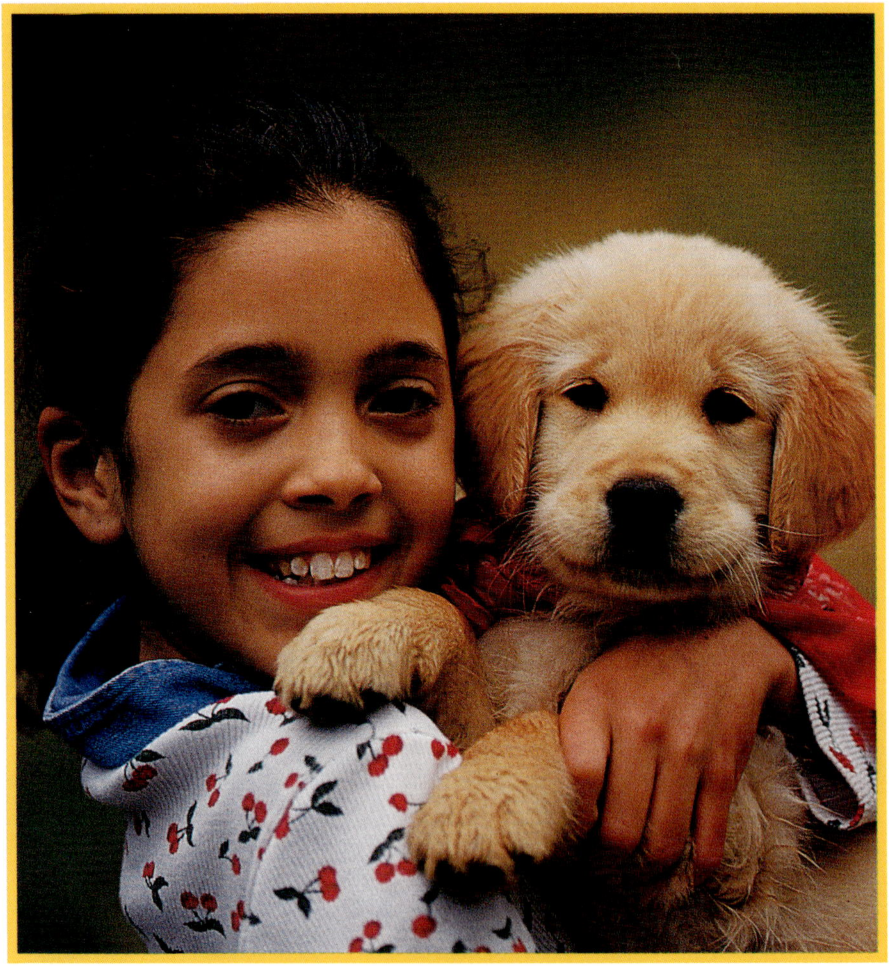

You can hug a pup.

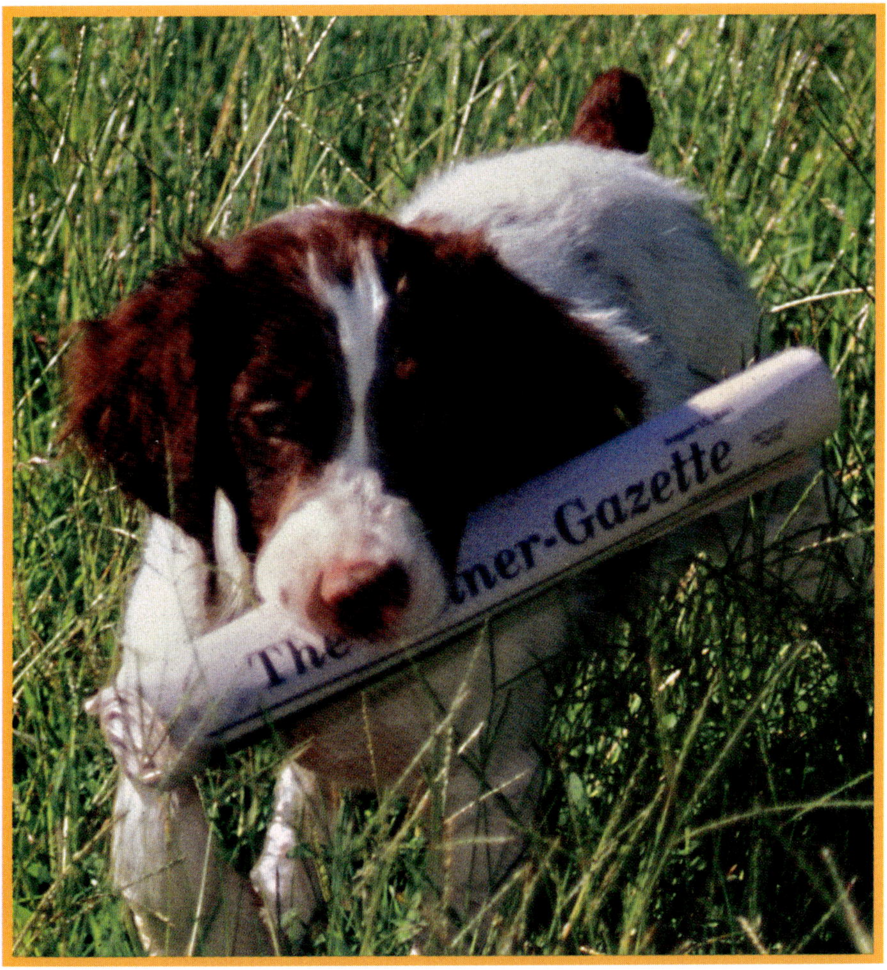

A pup can help you.

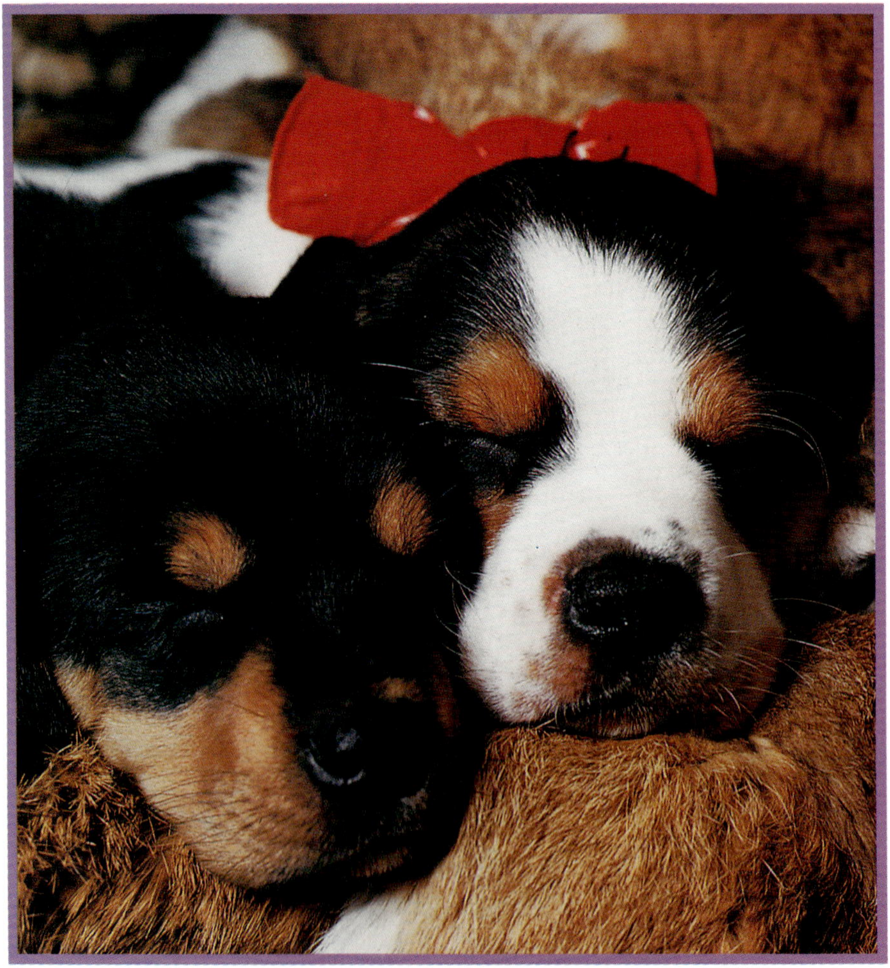

A pup can nap.

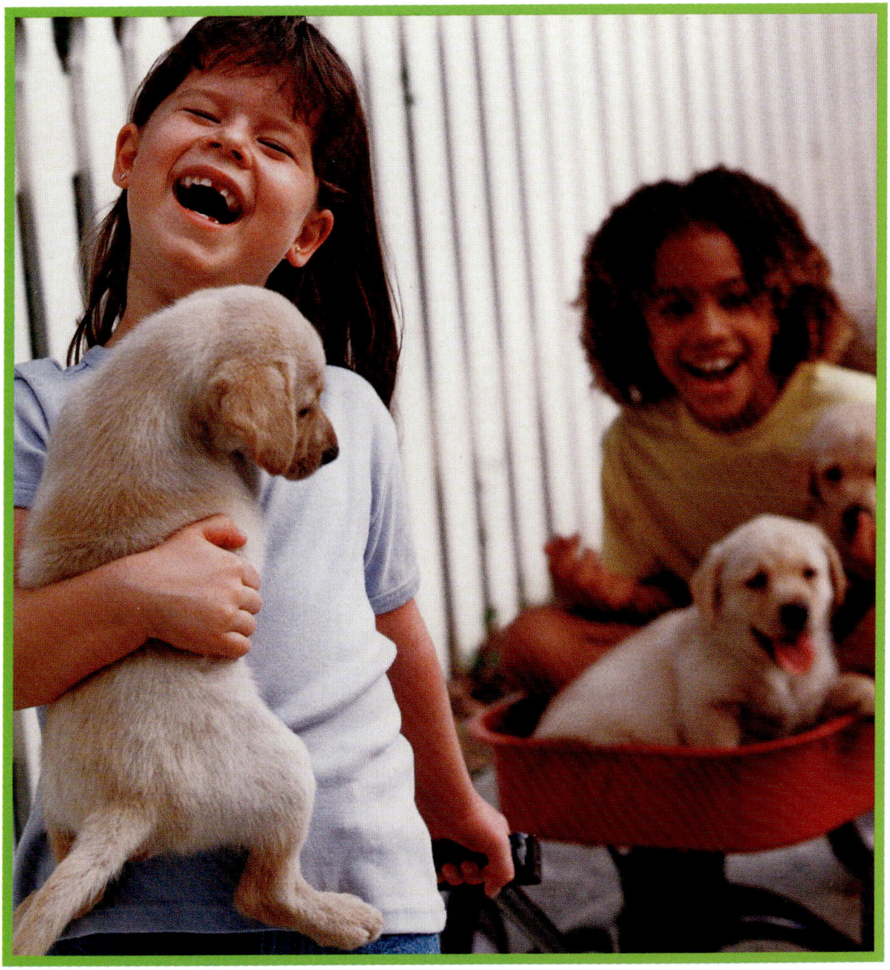

Is a pup fun?

Yes! Yes!

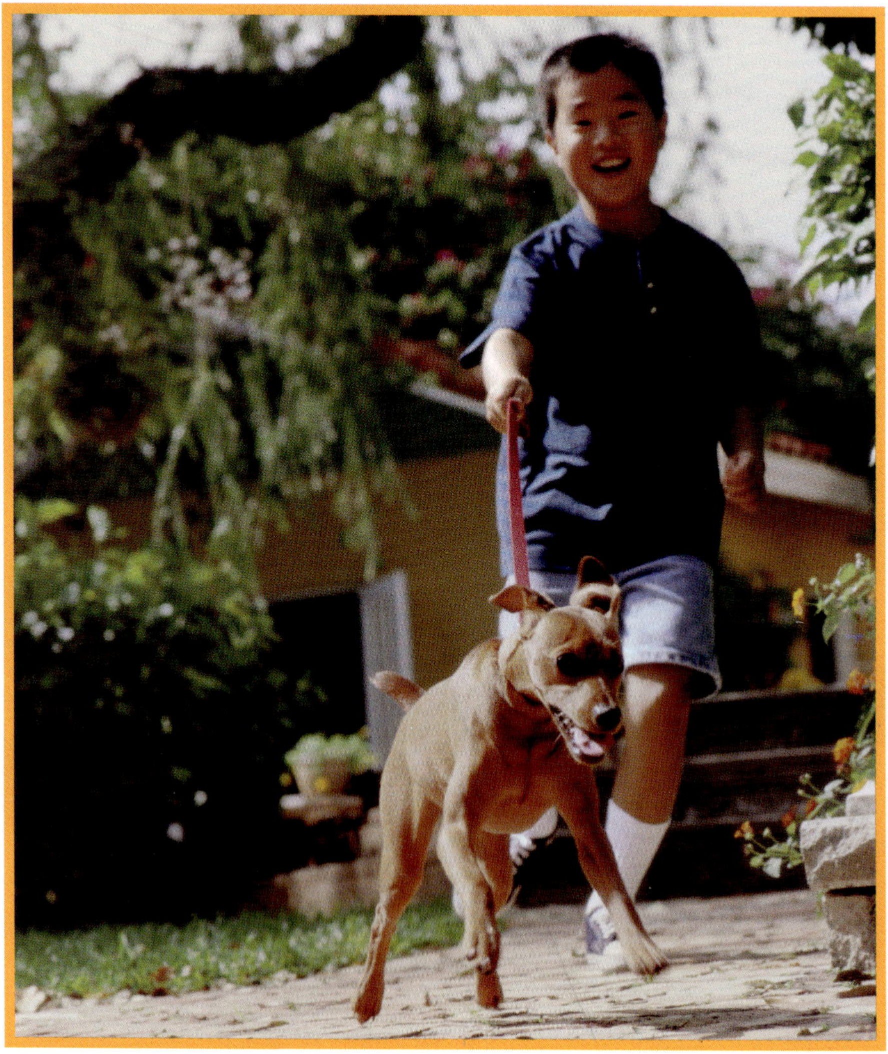

A pup is the best pet.